TO THE TOP OF THE WORLD

WRITTEN BY

Deborah Blumenthal

ILLUSTRATED BY

Anastasia Magloire Williams

Atheneum Books for Young Readers
NEW YORK · AMSTERDAM/ANTWERP · LONDON
TORONTO · SYDNEY/MELBOURNE · NEW DELHI

Barbara Hillary lived in a world of
you can, not *you can't*.

Barbara grew up poor
in Harlem, New York.
Her dad died when she was small,
and her single mom
earned whatever she could
by cleaning other people's homes.

But hardship didn't stop Barbara's mom
from teaching her
to have rich thoughts
and to dream big.

Barbara held on to her dreams,
and after working for fifty-five long years as a nurse,
she was ready
to explore exotic worlds she only knew about
from the adventure books she'd lost herself in
while growing up—
even if those books didn't include
heroes who looked like her.

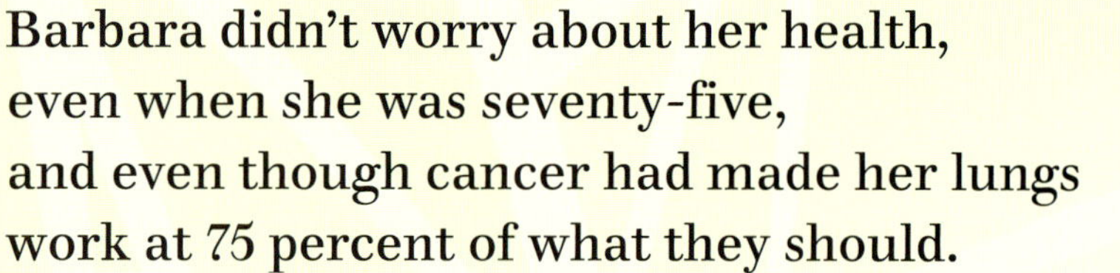

Barbara didn't worry about her health,
even when she was seventy-five,
and even though cancer had made her lungs
work at 75 percent of what they should.

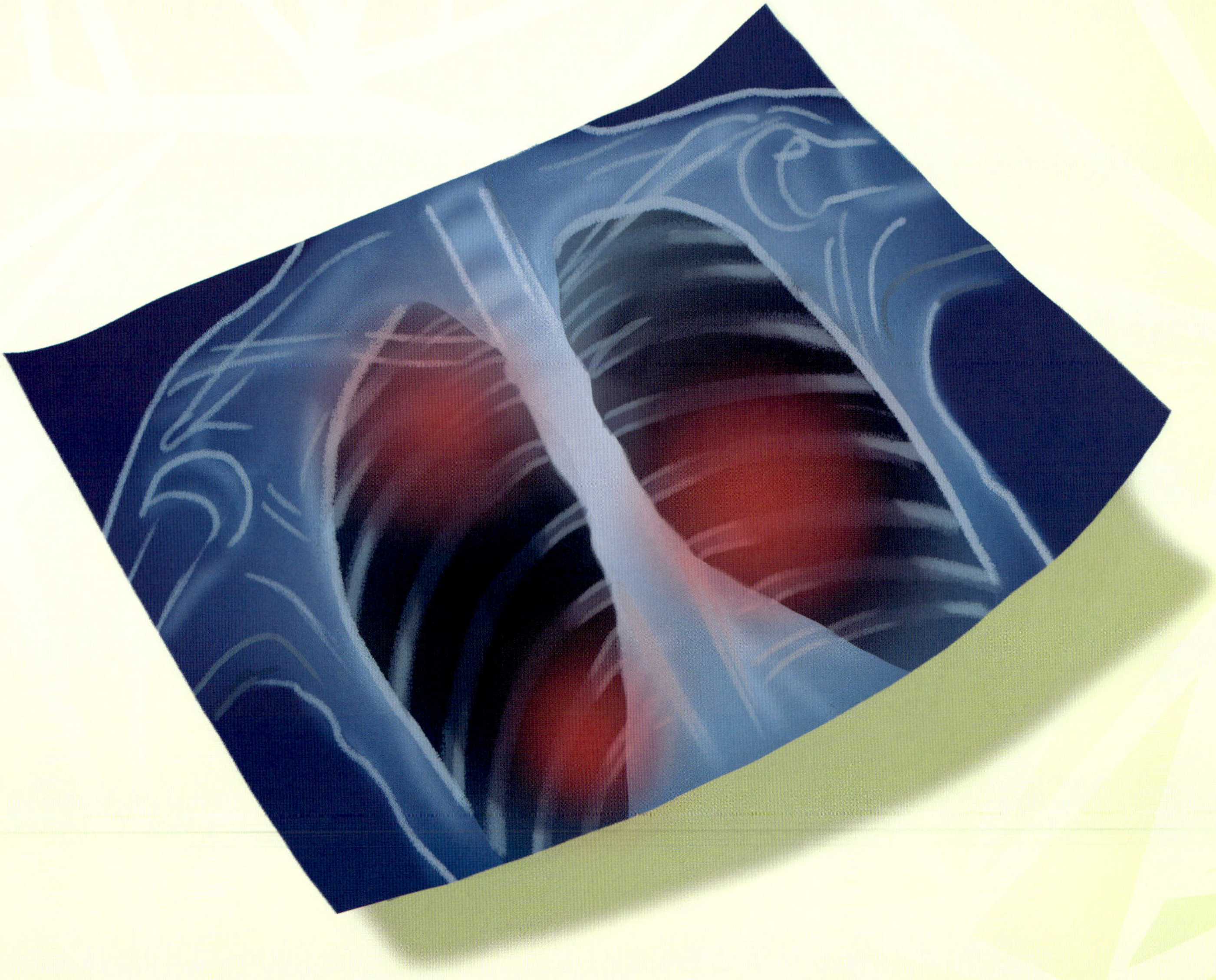

And she didn't worry about being too old,
because she dreamed of living,
not dying,
in a world of ***you can***,
not *you can't*.

Barbara loved snowy places,
so her first big dream was
to go dogsledding in Québec.
But first she'd have to learn how!
So off she went to Minnesota.

On her first run,
Barbara yelled, “Mush,”
and WHOA!—the dogs took off!
Then . . . WHOA! . . . BUMP! . . .
the sled hit a rock!

Barbara flew up into the air
and landed with a *BOOM*!

Barbara went home on crutches.
But that didn’t stop her.

Next, she traveled to Manitoba
to photograph polar bears,
who taught her to keep her distance.

Then Barbara dreamed bigger.

Why not go to the top of the world?

Barbara read about Matthew Henson,
the first Black man
to reach the North Pole.
When she realized that a Black woman
had never made the same trek,
an idea began to form.

She would be the adventurer
she never saw in her childhood books.

But a trip to the North Pole is expensive,
and she didn't have much money.
Plus, it would require muscle power,
which she'd have to build.

So she thought about
what she could do, and
not what stood in her way.

Barbara wrote letters again and again
asking companies for money.
Mostly, she was ignored.

But she didn't give up.

She ate more fruits and vegetables.
She took vitamins.
She lifted heavy weights at the gym,
and huffed and puffed on a treadmill
to grow strong enough
to cross-country ski in the icy cold.

While other people
were sunbathing at the beach,
Barbara dragged a sled
with a sandbag on top of it
along the sand because she'd have to ski
while pulling a heavy pack.

But *CRUUUNCH*! The sled collapsed under the weight!

Still, she didn't give up.

Barbara took skiing lessons
because to get to the top of the world,
you can't take a car;
you can't take a bus;
and you certainly can't take your bike or scooter.
The easiest way is to fly by helicopter.
But real adventurers go on cross-country skis—
traveling eight to ten hours each day
in the frigid cold.

And Barbara was an adventurer . . .

an adventurer who didn’t listen
to the people who told her
she was too old
or that she’d be eaten up by polar bears.

Because Barbara Hillary lived in a world of ***you can***,
not *you can’t*.

Barbara eventually raised the money
and flew to Norway.

Before the travel company would take her to the pole,
she had to pass a fitness test.
And thanks to her hard training, she did!

Next, she boarded a helicopter to the base camp Barneo,
about thirty miles from the North Pole.
Dozens of heated tents made up the camp,
all sitting on top of a giant mass of floating ice.

There was no running water,
and the sun shone night and day.

Barbara would stay at the base camp
until the weather was right,
so that another helicopter could carry her a bit farther.
And then the hardest part:
she'd ski for hours in the numbing cold
to reach the North Pole.

When the weather was finally clear,
Barbara boarded the helicopter
and made her way north.

In the brutal cold,
she and two guides
skied toward the pole,
pulling their heavy packs.

It felt like they would never get there.

They skied.

And skied.

And skied,
until hours later,
when one guide stopped,
looked around,
and turned to her.

"Barbara," he said,
taking in a deep breath.

"You're standing on top of the world."

It was April 23, 2007.
Barbara Hillary was seventy-five years old.

She had never experienced such sheer joy.
For the first few minutes,
she just screamed and screamed,
jumping up and down.

She was finally the adventurer she'd always read about.

In her excitement, she yanked off her gloves, holding up her thumbs for a photograph. She ended up with more than a picture—she also got frostbite!

For some people,
even breaking records,
breaking barriers,
and reaching the top of the world isn't enough.
Barbara wanted to keep going.

And she did.

Almost four years later,
on January 6, 2011, at age seventy-nine,
she made it to another remote part of the world,
a place even more bone-chilling:
the South Pole.

She was the first Black woman to go there, too.
This time, she brought milk chocolate to sweeten
her long journey.

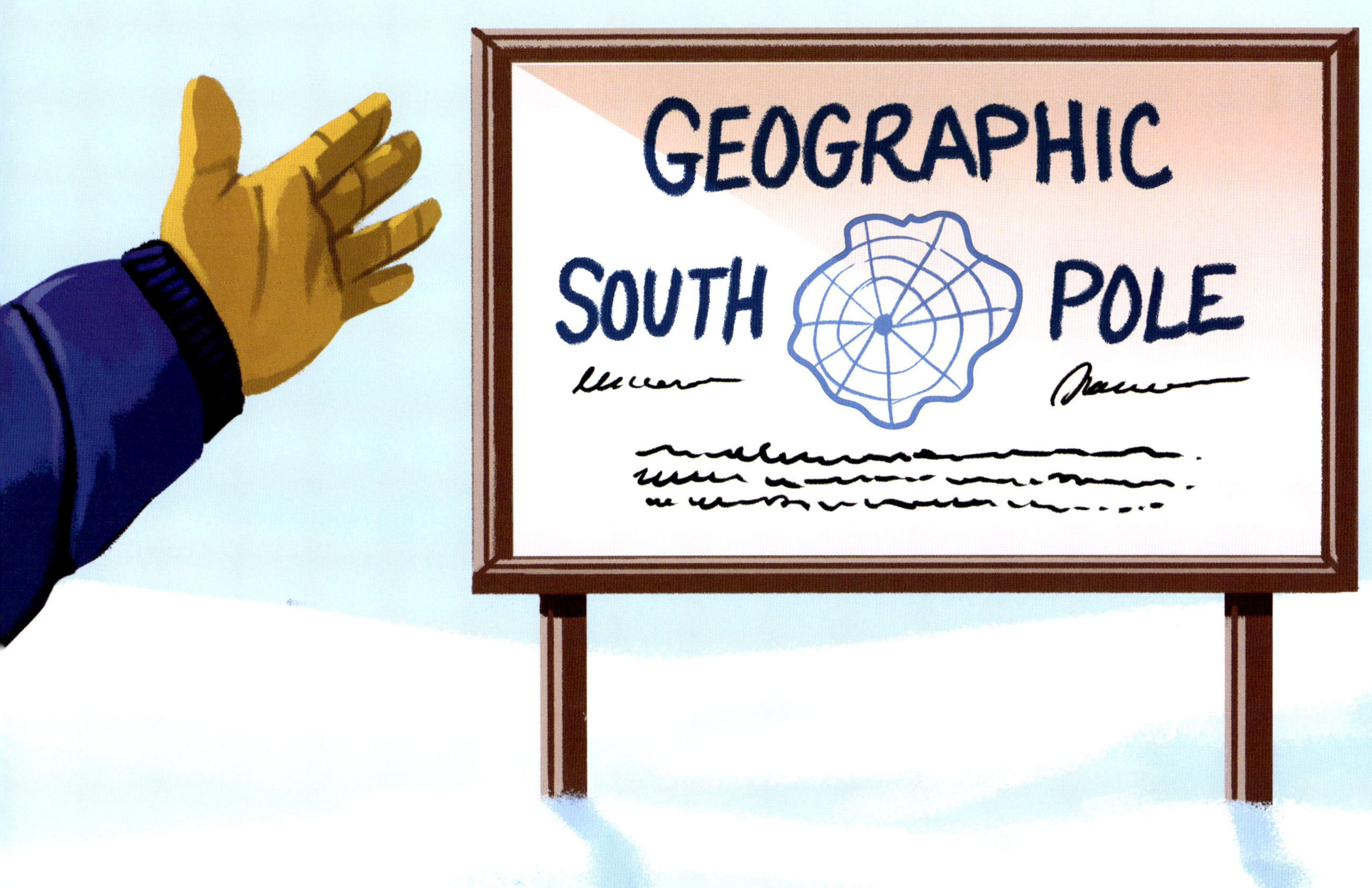

Barbara went on to visit other areas where most travelers don't go, learning about how climate change had harmed places like the poles, where temperatures were rising and melting the ice.

With global warming, the seas would get higher, flooding communities everywhere.

Some areas would be underwater;
others would be too hot.
Too cold.
Or too polluted.

If something wasn't done, no other adventurer would be able to reach those places again.

So along with traveling,
Barbara began to lecture, too.

Small things could make big differences:
Using wind and solar power instead of oil to heat homes.
Driving cars that save energy,
or biking instead of driving.
Eating more plant-based food.

When she was eighty-seven,
Barbara visited Outer Mongolia,
where she met nomadic people,
which means they move from place to place
and hunt for food
wherever they can find it.

Barbara asked these people—
who lived lives so different from her own—
how climate change was affecting them.

She would share their stories with the world.
She knew things could change.

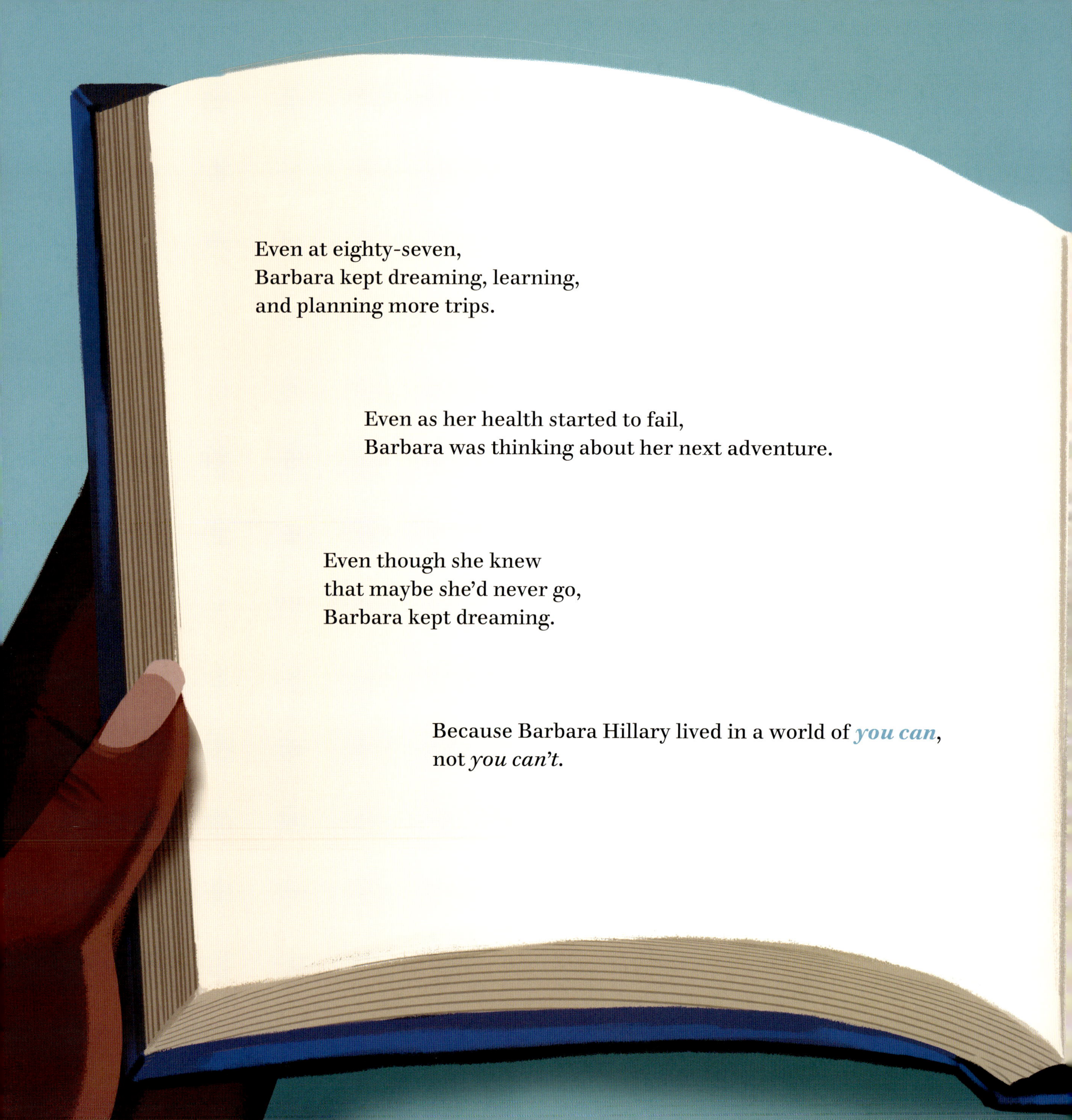

Even at eighty-seven,
Barbara kept dreaming, learning,
and planning more trips.

Even as her health started to fail,
Barbara was thinking about her next adventure.

Even though she knew
that maybe she'd never go,
Barbara kept dreaming.

Because Barbara Hillary lived in a world of ***you can***,
not *you can't*.

"Dreams," Barbara said,
"even if they don't come true, are important."

"I can close my eyes
and still see the wonderful mountains in Antarctica. . . .
They're blue-gray,
and there's the joy of silence."

AUTHOR'S NOTE

Barbara Hillary's passion was her ticket to unimaginable worlds that others only dream about. After retiring from a fifty-five-year career as a nurse—as well as a taxi driver and working in her community in Queens, New York (the same borough where I grew up)—she took classes in dogsledding. Despite getting thrown into the air when the dogs unexpectedly took off, Barbara put those classes to use in Québec and took photographs of polar bears in Manitoba, where she looked into the majestic creatures' eyes and saw herself as their next scrumptious meal.

Adventure was always in Barbara's blood, inspired perhaps by her love of reading books like *Robinson Crusoe* when she was young. So when Barbara heard that no Black woman had ever been to the North Pole, she knew what her next adventure would be. Her age—seventy-five—was only a number.

Having survived breast cancer when she was in her twenties and lung cancer when she was sixty-seven (leaving her with only three-quarters of her normal lung capacity), Barbara knew about surviving the odds. Still, she was all-in—living life to the fullest was worth the gamble. So she got in shape with a trainer and learned to ski, then signed on for an expedition to the North Pole with a group called Eagles Cry Adventurers. Barbara was flown to a Norwegian base camp, where she stayed until another helicopter dropped her into an area within hours of the North Pole on skis.

When asked if she had been afraid of sleeping in a tent with polar bears outside, Barbara said, yes, she was nervous . . . but not about the polar bears! "You pray you don't have to wee-wee," she said. "You haven't lived until you have to get dressed and go out when it's thirty below."

Reaching the North Pole would have been enough of an achievement for most people, but not Barbara Hillary. Almost four years later, she journeyed to the South Pole. Her accomplishments didn't end with that second journey, either. Concern with how climate change was affecting remote parts of the world led Barbara to become an environmental activist.

"Everything about her was fascinating, convention-breaking, and confounding," a friend told radio station 1010 WINS in an interview. "Her record-setting treks, her defeat over cancer, her arduous fight to get her house back after Hurricane Sandy." She appreciated life and had great stories to tell, Barbara's friend said. And oh, "the roses and miraculous tomatoes she grew."

I was drawn to find out more about Barbara Hillary after reading her obituary in the *New York Times*, shortly after she died on November 23, 2019, in the season of twenty-four-hour sun in the South Pole. A woman of seventy-five going to the North Pole? Then going to the

South Pole at seventy-nine? If there was ever a great story about human achievement, this was it.

While Barbara's accomplishments were extraordinary for a person of any age, what struck me most was the fact that she refused to let time, fear, or negativity get in the way of her dreams. In one article, she told a reporter that when she flew to Norway to begin her trek to the North Pole, she had a tear in a tendon in her shoulder, but she kept it to herself.

"I know when you focus on something negative, everyone is watching," she said. "This way no one had to give me any special treatment or make special provisions or think, 'Maybe she can't do it.'"

Barbara wanted her life to be an empowering example for others. She wanted to support those with cancer and other diseases. She wanted to help Black Americans, and women in particular, bolster their sense of self-worth. And I imagine that if she had been asked to give advice to children, she would have told them to follow their dreams, ignoring anyone or anything that stood in their way.

Many thanks to the *New Yorker* for permission to use quotations from Lauren Collins's interviews with Barbara Hillary.

Many thanks to Lou Dzierzak for permission to use quotations from his article "Barbara Hillary Triumph at the Top of the World, from Cross Country Skier."

To Ralph.

—D. B.

For those who worry they've missed their chance for a grand adventure—the day you begin, you'll be right on time!

—A. M. W.

atheneum ATHENEUM BOOKS FOR YOUNG READERS • An imprint of Simon & Schuster Children's Publishing Division • 1230 Avenue of the Americas, New York, New York 10020 • For more than 100 years, Simon & Schuster has championed authors and the stories they create. By respecting the copyright of an author's intellectual property, you enable Simon & Schuster and the author to continue publishing exceptional books for years to come. We thank you for supporting the author's copyright by purchasing an authorized edition of this book. No amount of this book may be reproduced or stored in any format, nor may it be uploaded to any website, database, language-learning model, or other repository, retrieval, or artificial intelligence system without express permission. All rights reserved. Inquiries may be directed to Simon & Schuster, 1230 Avenue of the Americas, New York, NY 10020 or permissions@simonandschuster.com. • Text © 2025 by Deborah Blumenthal • Illustration © 2025 by Anastasia Magloire Williams • All rights reserved, including the right of reproduction in whole or in part in any form. • ATHENEUM BOOKS FOR YOUNG READERS is a registered trademark of Simon & Schuster, LLC. Atheneum logo is a trademark of Simon & Schuster, LLC. • For information about special discounts for bulk purchases, please contact Simon & Schuster Special Sales at 1-866-506-1949 or business@simonandschuster.com. • Simon & Schuster strongly believes in freedom of expression and stands against censorship in all its forms. For more information, visit BooksBelong.com. • The Simon & Schuster Speakers Bureau can bring authors to your live event. For more information or to book an event, contact the Simon & Schuster Speakers Bureau at 1-866-248-3049 or visit our website at www.simonspeakers.com. • The text for this book was set in Walbaum. • The illustrations for this book were rendered digitally. • Manufactured in China • 0625 SCP • First Edition • 10 9 8 7 6 5 4 3 2 1 • Library of Congress Cataloging-in-Publication Data • Names: Blumenthal, Deborah, author. | Williams, Anastasia Magloire, illustrator. • Title: To the top of the world : Barbara Hillary, the first Black woman to reach the North and South Poles / Deborah Blumenthal ; illustrated by Anastasia Magloire Williams. • Description: First edition. | New York : Atheneum Books for Young Readers, 2025. | Audience: Ages 4–8 | Audience: Grades 2–3 | Summary: "A nonfiction picture book about Barbara Hillary, the first Black woman to reach both the North and South Poles"—Provided by publisher. • Identifiers: LCCN 2023025214 (print) | LCCN 2023025215 (ebook) | ISBN 9781665927734 (hardcover) | ISBN 9781665927741 (ebook) • Subjects: LCSH: Hillary, Barbara, 1931–2019—Juvenile literature. | North Pole—Juvenile literature. | South Pole—Juvenile literature. | Arctic regions—Discovery and exploration—American—Juvenile literature. | Antarctica—Discovery and exploration—American—Juvenile literature. | African American explorers—Biography—Juvenile literature. | Women explorers—United States—Biography—Juvenile literature. • Classification: LCC G585.H55 B58 2025 (print) | LCC G585.H55 (ebook) | DDC 910.92 [B]—dc23/eng/20231120 • LC record available at https://lccn.loc.gov/2023025214 • LC ebook record available at https://lccn.loc.gov/2023025215